Time Travel

If Possible ?

Julbiharahamed K

Title: Time Travel : If Possible?

Author : Julbiharahamed K

Edition : 1st Edition

Year : 2017

ISBN-13: 9781549758836
ISBN-10: 1549758836

CONTENTS

INTRODUCTION

Time Travel is a desire for all of us. In our life we are all think about our old memories and desire about the future. It's just not to say no. So when we think of it, how many of us remember when we go to that time? Is it possible?

In this book I will tell you about this time travel. Let's look at the possibilities and the scientific explanation.

We can also see the contradiction and the evidence that it has already happened.

TIME AND TIME TRAVEL

In recent days, most of us have often heard words like time travel time machine and etc., in science fiction stories and movies. One of those stories goes forwards or backwards in time0 with the help of a time machine. And in some of the stories one would accidentally in time travel.

If you can overcome the time travel like these stories, the opportunities for it are very low, and there are opportunities to prove it. Let's see a little bit about what time is ahead of us to see about the time.

Most of us thinking time is just an unit. Measurements such as how much time a person spends. But the best term for time is a dimension. The Albert Einstein thought *"There are really four dimensions, three which we call the three planes of Space, and fourth, Time"*. According to physics, time is a predecessor from the present time to the random area of the future.

Time and Distance

Relationship between time and time is very complex. Between the two places (or the objects) is the distance and not only the distance time also. For example the distance between two places are 200 miles. Traveling time on car almost 5 hour, But the same distance on flight just 1 to 1 and half hours only. The equation for time is,

Speed = Distance / Time

Time = Distance / Speed

Although the concept of the term has long been considered, but the modern ideas appeared from **H.G.WELLS**'s book **Time Machine (1895)** (H.G. Wells is another dimension 10 years before the publication of general relativity policies in Time Machine) *H.G.Wells* and *Jules Verne* are known as the father of science fiction stories. As we know H.G.Wells is the first person to write about the time travel. He has written in his book 10 years before Einstein. Likewise, in 1921, **Emmett J. Flynn's** film, **Connecticut Yankee in King Arthur's Court**, was the first film about the time travel.

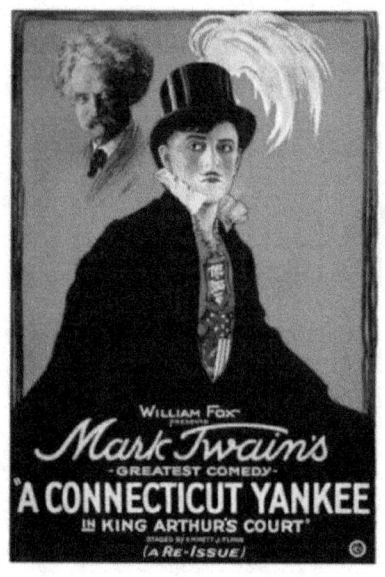

According to Einstein's theory of relativity, an object that accelerate parallel to the speed of light (C = 3 × 10⁸ m / s) passes over time. If you want to accelerate the object to a velocity of light, you must go at speeds of 3,00,000 Km per second. At this pace we can round our earth in just 8 seconds. Practically it's impossible but theoretically it's possible.

Time travel is a change in time between different points of time. Before talking about time travel. We need to know about the Albert Einstein Relativity Theory and the time travel principle of Stephen Hawking.

Relativity

A **Theory Of Relativity** is a physical philosophy brought by the German scientist Albert Einstein. These came in two forms: **Special Theory of Relativity** (1905) and **General Theory of Relativity** (1916). Alfred Einstein's philosophy was given a great reputation to him. Then he was compared to legends Nikolas Copernicus, Kepler and Isaac

Newton . It is still a milestone in physics history

The theory of relativity puts forward some ideas. They,

1. Different sizes calculated on will depend on the velocity of the viewer. In particular, the time and the space may be shorter. That is, all the quantities that we calculate will depend on our speed. If we go fast we will see less and slowly going to be over. The time and the space may be expanded or shorter.

2. **Space Time:** The time and space should be taken into account. The other will depend on one. The spaces and the time must be taken into account. Its actions will depend on one another.

3. The speed of light for any observer is constant.

And Einstein made it clear that even the smallest particles could release a mass of energy through its equation $E = MC^2$.

Hawking Principle

The time travel policy of Stephen Hawking, an English-based physicist, is famous and is a universally accepted philosophy. They pass the time when a material crosses the gravitational fields such as **Black Hole** and **Worm Hole**.

This is a hypothetical principle, which means that we have a very large spacecraft capable of running for many years, They reach

50% of the speed of light in 2 years, and in 2 years it will reach 90% of the speed of light, even if 2 years (i.e. 6 years total) 98% of the speed of light. They will last for six years but at the same time the earth will last 80 years. There are a few other theories besides that.

Types

The Time Travel is two types of stories and scientific concepts. **Forward Time Travel** and **Backward Time Travel** both seem to be same, but theoretically different.

Time Machine is a hypothetical tool for a Time Travel. In this case, only in stories. They are built in many forms of time machine. A small hand clock has been built in many variants, starting with a very large car. Many scientists tried and failed in this attempt.

FORWARD TIME TRAVEL

Forward **Time Travel** or **Time Travel To The Future** is a possible reaction, According to Relativity Theories. We will be saw on science fiction stories and films, A person with machine or portal help of traveling for time. In short, a trip for present to future.

According to Einstein's general theory of relativity, when the object is accelerated at the speed of light, the material passes over time. The speed of light is 3,00,000 KM per second, No thing to achieve the speed of light. The equivalent speed of light is 0.95 C or 0.99 C (The fastest man-made vehicle speed is just 58,000 KM).

Before we see this we will see a small event.

An island crash is taking place, and after a few hours the fire is still in the same area. If both of these events happened in the same place, we will say yes. But, The same question in space someone asks his response of course **Not** only thing that used to say. It's how to the same event two other places in that the ask. The reason is the Earth's rotation's.

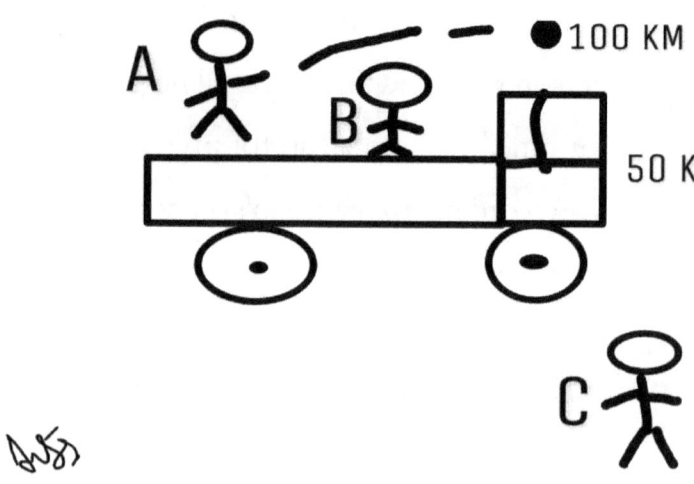

Similarly, a small example, 50 km speed will be a vehicle, the speed of the ball is 100 KM for a person with a ball in a 100 KM speed, but its speed is 150 KM.

Einstein has given the explanation for these two events in its special relativity, *where the moving clock speed is less than the speed of the stable clock. How is it going to change the speed of the clock?*

For example, one of the twins born together is about one in the speed of light, about 10 years into space. At the age of 40, the earth is 80 years old. Is this possible? Because of this, Einstein says it's Earth's cycle. Yes, according to his theory of relativity, a clock that runs on Earth, a stable clock in space.

You may have a question. Similarly, there is a change in the car and the change in the car. We could not feel it because it was less than nano seconds.

Directed by Christopher Nolan in the film Interstellar. The story of this film is whether the hero in this film is to go to a space to find a resident other than the earth and find out the problems that are going back and return to the earth. When a hero goes into space, his daughter is 10 years old when he returns from space after a few months and his daughter is 70. But his age is exactly the same.

In another scene, except for one of the traveling crews, the others will go to a new planet. They will only be there for a few hours. But there will be more years of age than the age of the partner.

In 1971, a test was conducted to prove this relativity. 5 synchronized atomic clock and put one of them in a space station and 4 clocks on a passenger plane. The plane was traveled twice earth in a steady speed. At the end of the test, 4 bikes on the plane ran less than 40 nanoseconds more than the clock in space. The clock to which the clock runs, rather than a constant clock, has proved to be a special relativity.

Let's now come to the speed of light. Can it really go through the speed of light? How can it be possible? We can not be sure of the technology that we have now.

Can one be able to cross the board with the help of a portal? Scientists say. But they have named it **Worm Hole**.

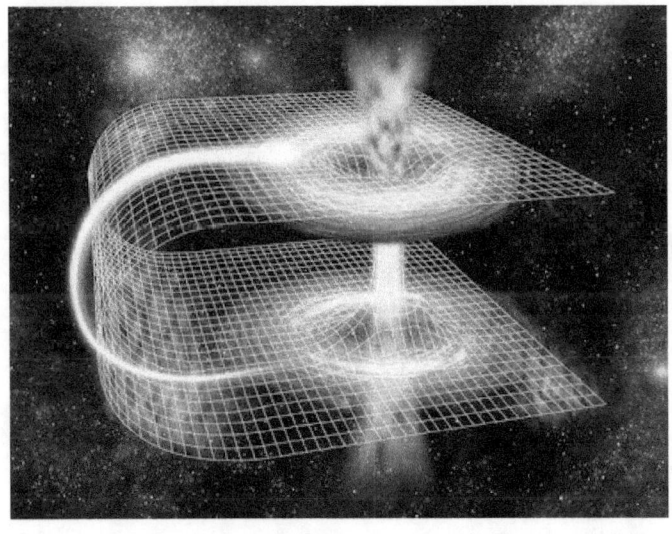

A hypothetical path accepted by scientist is the best way to pass the time according to the theories of relativity. By Einstein's general relativity, they will link two other galaxies that are multi billion light years away. It is also sometimes known as a **Transporter**.

When it comes to describing the nature of the hockey period, it is said to be expanding and contracting. That is, the period is not unchanging. By this nature it will go forward or backward. For instance, he says that when we look at the other part from a very heavy weight, the other area will appear to be higher.

BACKWARD TIME TRAVEL

All have desire for travelling back in time.

Scientist say it's backward time travel or negative time travel. Einstein's Theory of Relativity and Hawking's Explanation about Black Hole are alone to explain backward time travel. According to Einstein's theory of relativity, the material goes back in time at a higher speed than the speed of light.

We saw in the last part that the speed of light was not possible, so the backdrop was a practice that was not practiced in practice, but possible in philosophy.

Before we can see this, we see a lot of actions that occur at the speed of light. The speed of light is the constant, which means it does not decrease. Its speed is 3 lakh kilometers (1 lakh 84 thousand miles) per second. If we go at this pace, we can see the embedded objects appear to be like an idol. We can not hear any sounds, and we can not feel any color. And everything is dark.

Perhaps one can go backwards in the direction of speed rather than the speed of light. In story telling in the stories, the backdated time cycle is going backwards. Researchers claim that this is not possible in practical science.

Stephen Hawking says two events are possible. The first known **Black Hole** , which is the most densely vacuum in the center of the galaxy. It is a giant animal that swallows all the stars and planets. In his book A Brief History Of Time, he says that the universe will appear in the Big Bang and end with Black Hole. He says this is a way to get back to the time travel.

Although it is practically impossible to use it is possible on the basis of policy. During a period of time when one traveled through a black hole, one of the fastest passengers on an ocean cruise shielded and gravitated by its attractiveness. In the opposite direction (backwards).

The retreat Time travel or past tense may be theory of theoretical methods using some methods. Cosmic strings or using black holes to travel. One light than the speed of fast information or pass an object, if possible, the information or material According to special relativity back in time at different locations "*at the same time*," the two events happened whether in some cases different reference frames will disagree, as well as the two event Disagree on the order of thousands. In both cases one sends the information (Sender), another information receives (Receiver). However, the hypothetical signal faster than light in the city before it was sent the signal was received, some frames will remain, the signal moved backwards in time, you can say. While special relativity does not impede the ability to accelerate the speed of light at the speed of light, and relativity does not prevent the theoretical possibility of all-time fast-moving at all times, the quantum . The possibility of using these to change is actually impossible and it is noteworthy.

EVIDENCES

Do you have any evidence or traces?

Not directly, but indirectly has many of our history. Let's see some of them in the region.

As far as our ancestors are concerned, they use religion as an instrument that brings many ideas to the people. There are some tips on them.

Mihraj

The journey of Mihraj is Prophet Mohammed's Prophet was taken from the city of Makkah in the Arabian desert to the Al-Aqsa

Mosque in Jerusalem by the night of the angel Gabriel (the Heavens). ISRAA (in Arabic - to take the night). The mere pilgrimage from the Al-Aqsa mosque, known as the Pyatul Muktus, is called Mihraj.

One of the details of this is the 1486 KM of Jerusalem from Mecca. Then he went from there to the celestial bulak. In this we have to look for a burac vehicle that is like a wild beast. This event was the year 621 AD. It may have been the same at that time.

Seven Sleepers

This event is mentioned in Islam and Christianity. In the 18th chapter of the Qur'an (Al Qafu - Caves) and the Bible is in Seven Sleepers.

It is said that a group of young people go into a cave to escape from the thugs. There they spend a night out to escape. But they come out after 300 years.

King Raivada

In Hindu mythology, stories about Kakudmi are coming. The king of the Suryakya dynasty is a woman named Revathi. Gudhudmi, who feels that nobody is worthy of marrying a beautiful woman like the angel, is going to consult Revathy and consult with Brahma in the Brahmava.

Time in the universe is not the same for everyone. 108 years have passed since you were here. It will be millions of years. Brahma, who is awake in what to do, says, "Lord Vishnu has now taken the form of Krishna and bala ram many others on earth. Many of them will be like your girl. " When the Kakudmi returns to the earth he enters the future.

One equal to hundred

Buddha has said in one of the teachings mentioned in Buddhism. The Buddha is here to spend hundreds of years on his disciples, just a year in paradise.

This is much more likely to be accompanied by the time difference we've seen in the time forward.

Yokashima Taro

The Japanese story, Yokashima Taro, is a well-known storytelling associated with traveling in the ancient times. It is a story about a young fisherman named Yokashima Taro, who goes to the deep-royal palace three days. After returning to his home in his town, he finds the future for the next three years, where he has long been forgotten and his house was destroyed and his family had already died. Another distinguishing example of this type of story in Dalmaton is the Honey Hame'Agel story, which lasts 70 years, and when he wakes up, his grandchildren see his grandparents as his grandparents and his friends and family.

Another such story is Rip Van Wing. It's also like Dorothy. In the story of the story of Rip Van Wing, he drank wine one day and slept in a wooded hill. Then he got up and turned around. Everybody is older. He has a long lasting and shimmer skin. Then he feels he's been twenty years old.

Furthermore, ancient folk tales and myths were sometimes interested in traveling forward.

Modern Evidence

In this, we will see randomly about the timing of some photos and video sources available to us.

The two pictures that were built above the place taken in the last century, unnamed photos.

Let's see the first picture. It looks like something like a ceremony. Look at the difference in everyone except the person who is hovering around. Everyone's style is in old ways. But his dress, hair dressing. It seems like the modern method is bleeding.

In the second picture is almost like that. Everybody's clothing has a great deal of change except the one mentioned.

The next picture is a coincidental picture taken in the Mike Tyson Boxing Tournament. One fan of the film is taking the film. But it looks like the latest smart phone in the camera.

The world's first camera phone came in 2000. So how could one use it in the 1980s?

Some people claim that it is a popular Casio Camera at that time. But this can be done in the crosswalk. But the person takes the picture in the long waist.

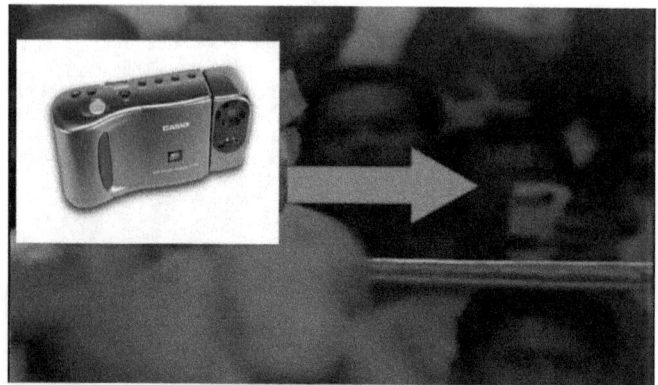

Hakan Nordkvist is famous for a few years ago. He took video himself at old the video and posted it on the Internet.

It was seen by millions of people. The source has shown both the hands and the tattoos.

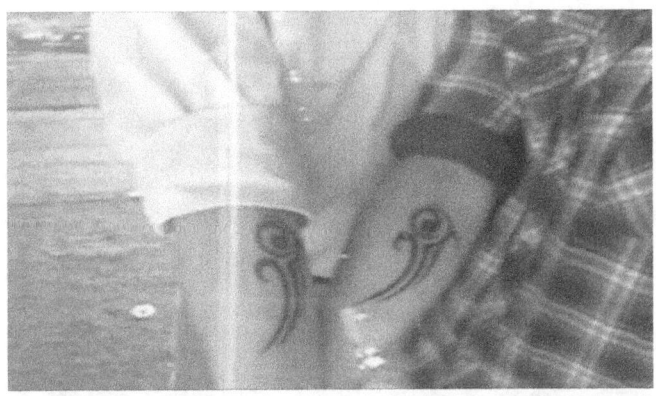

Charlie Chaplin's film The Circus, released in 1928, will feature a person's hand talking to a scene. Only 50 years later the hand was found.

In another picture like this, a woman is talking in a sense almost like a man talking.

Next evidence such a great evidence to proof time travel. In this pictures one person theft using time travel.

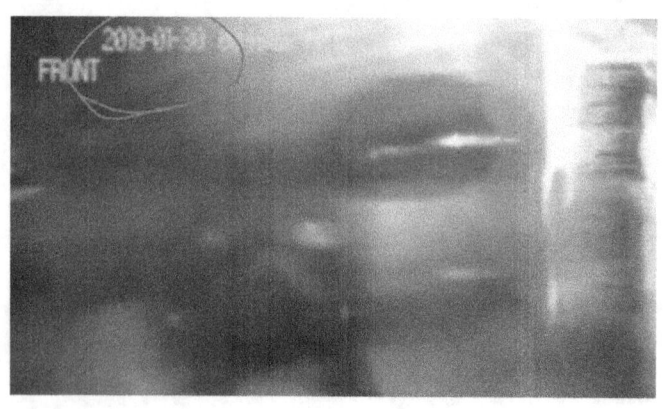

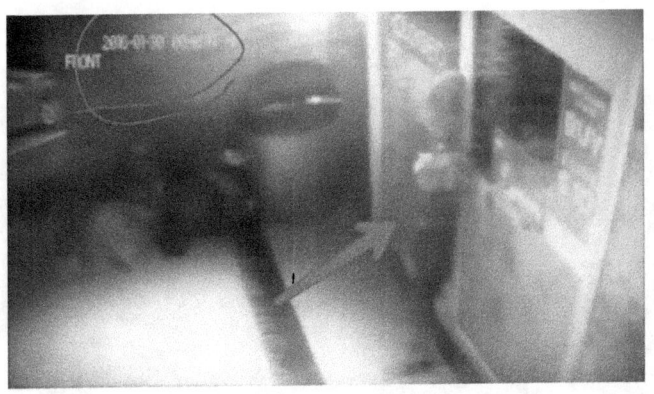

Take a look at the pictures above. It has been recorded on a surveillance camera. On January 30, 2016, a date is passed through the door. He uses the mystical algorithm like portal. At that time, the date of the camera is changed to 30 January 2019.

This pictures is on the road surveillance camera in Japan. In this, there is an accident on a truck bike. At that time a mysterious person comes along with the vehicle. In the last movie, the truck's driver looks stunning.

We can not say for sure that it is possible to keep the trip possible. But if you look at the next pictures, it seems that the trip is about.

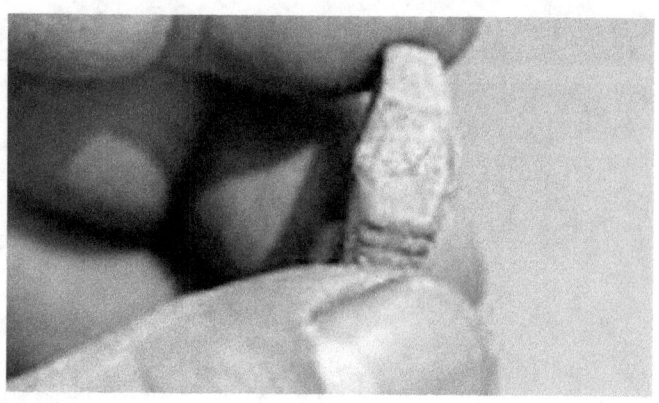

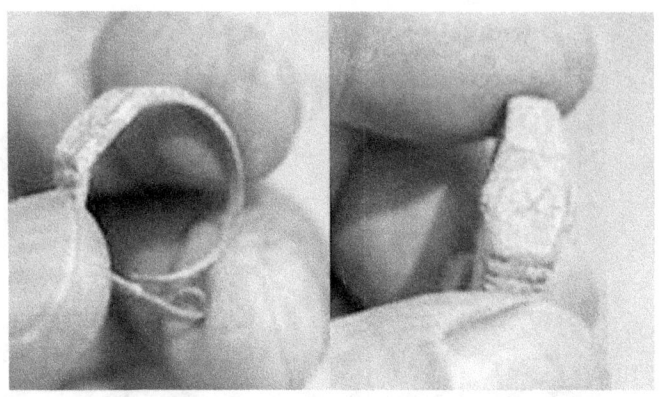

The pictures on the top were obtained by excavations in China in 2008. It got a coffin. This is like a clock ring. Analysts claim it is 800 years old.

How is it possible?

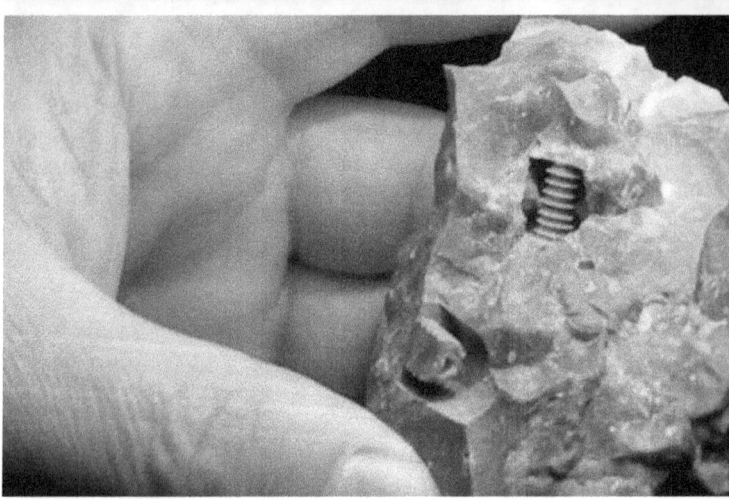

In the same way as another, it is a signal and screw cut by a modern machine.

The sculpture in the church is in a church in the French. That figure is like an astronaut. It was designed in the 1600s.

We can only conclude that we can only make the trip possible with the evidence we have seen.

TIME TRAVEL PARADOXES

The last few chapters we saw about time travel and their possibilities, Evidences. But on this chapter you think time travel is not possible. Because, time travel theory have lot of Paradox.

Grandfather Paradox

Grandfather Paradox is a hypothetical explanation for time travel is not possible. On this theory one person build a time machine. He travelling back in time and kill their own grandfather their father/mother borned (If Accidentally or Pre Planned).

His father would not have been born if he killed his grandfather. He could not have been born. How can someone who does not have a time machine build up.

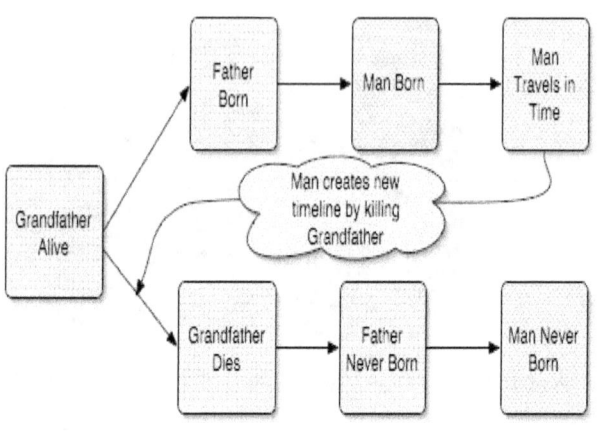

Nathaniel Schachner is a first person used this theory on his book *Ancestral Voices (1933)*. The scientist still do not have complete description about this paradox. David Deutsch (Quantum Scientist, Israel) have another clarification about this. He use about that **'Another Universe'**. He thought Time Travel environment create the **'Another Universe'**.

Hitler Murder Paradox

This is similar to grandfather paradox but different level. That time the traveler has already happened and something remarkably goes back to time to change. Unlike the Grandfather Paradox (which we assume would self-correct despite our best efforts), the change that one wishes to affect in the Hitler's Murder Paradox is one that is more technically feasible - as in not intrinsically paradoxical - but still ultimately problematic. the name comes from the idea that one could theoretically go back in time and kill adolf hitler before the holocaust happened, To lead to some kind of downward spiraling domino effect with plenty of other consequences that the well-intentioned time traveler probably didn't consider, and which ultimately might lead to a worse situation than that which the time traveler had hoped to prevent.

Butterfly Effect

Similar to the cascading domino effect of the Hitler's Murder Paradox, but on a different level. Whereas killing Hitler would obviously be a landmark event with quite a significant historical impact, something like, say, accidentally stepping on a bug in the past probably wouldn't have as big of an effect, right?

Have you even been paying attention? Of course it will! That's the whole point of a time travel paradox! Just like the way that a butterfly flapping its wings in Brazil can affect a weather system in Texas, one tiny change in the past can lead to all kinds of Rube Goldbergian complications that can subtly -- or seriously -- affect the present. The term "Butterfly Effect" is actually derived from "A Sound of Thunder," a short story by Ray Bradbury, in which a character accidentally steps on a butterfly in prehistoric times and causes catastrophic changes in the future from which he came.

For example, In Orpheus With Clay Feet by Philip K. Dick, the main character, Jesse Slade, enlists in the services of a time travel tourism agency, who set him up with a trip that allows him to go back in time and act as a muse for some significant historical figure. Slade chooses to go back and inspire his favorite science fiction writer Jack Dowland (which was also Dick's pen name). Unfortunately, in his efforts to inspire Dowland's monumental science fiction work, Slade directly reveals to Dowland that he is a time traveler hoping to inspire his work. Dowland takes this as an insulting ruse, and as a result, never becomes the great science fiction writer that he is meant to be. He does, however, publish a single science short story, under the pen name Philip K. Dick: a story called Orpheus With Clay Feet, about a time traveler that goes back in time to inspire his favorite science fiction writer, a man named Jack Dowland.

Ontological Paradox

Ontological paradox arises when a person or object is sent through time and recovered by another person, whose actions then lead to the original person or object back to the time from when it came in the first place, thus creating an endless loop with no discernible point of origin. Thus, the original person or object is essentially "pulling itself up by its own bootstraps," hence the nickname (thanks in no small part to the Robert Heinlein story "By His Bootstraps").

The Terminator films are a prime and popular example of the Ontological Paradox. In the future, a Terminator is sent back in time to kill the mother of resistance leader John Connor before he is born. While the original T-800 is ultimately destroyed, the leftover pieces are found by scientists who use the technological to...develop and create Skynet, and the Terminator-series robots. Skynet would have never been created if Skynet hadn't taken over the world and then sent a Terminator back in time to get destroyed and ultimately lead to the creation of Skynet.

There's also the fact that Future John Connor sends his buddy Kyle Reese back in time to protect his mother from the T-800, only Kyle ends up totally bangin' John's mom (dude high five! I mean, not cool, man) and impregnates her with his buddy John Connor. So to top it all off, if John hadn't sent his friend back in time, his friend would never have had sex with John's mom, and John would never have been born.

In the Terminator films are a prime and popular example of the Ontological Paradox. In the future, a Terminator is sent back in time to kill the mother of resistance leader John Connor before he is born. While the original

T-800 is ultimately destroyed, the leftover pieces are found by scientists who use the technological to...develop and create Skynet, and the Terminator-series robots. Skynet would have never been created if Skynet hadn't taken over the world and then sent a Terminator back in time to get destroyed and ultimately lead to the creation of Skynet.

There's also the fact that Future John Connor sends his buddy Kyle Reese back in time to protect his mother from the T-800, only Kyle ends up totally bangin' John's mom (dude high five! I mean, not cool, man) and impregnates her with his buddy John Connor. So to top it all off, if John hadn't sent his friend back in time, his friend would never have had sex with John's mom, and John would never have been born

Self Visitation Paradox

One of the reasons for this is that it is the center of attention that opponents of these time travel policies are convinced that this is impossible.

According to this, a young man has a method of creating an elderly time machine. He uses the method to model his time machine. Using the results of the games, using the results of the stock markets, he becomes aware of the results. He could not find the old man anywhere.

He then went back to his old age and taught himself the method of designing the time machine. This is the young man who was previously old. Question is, Who designed a time machine? (The event that itself meets itself in science fiction stories and movies is shown to be dangerous and in some cases help).

This is also known as **_Self Visitation Paradox_**.

The Bootstrap Paradox exists when a chain of cause-effect events is circular. Event A causes Event B, which causes Event C, which causes Event A. Is your head spinning?

Imagine that your kooky great-uncle passes away one day, and in his will, he decides to leave you his diary. He was a treasure seeker in his youth, and amid tales of high adventure, you find mention of one peculiar treasure he uncovered in a strange, metallic box - blueprints for a time machine. He had the decency to include the folded-up blueprints within the diary, and as an esteemed physicist, you recognize that the science is sound and set out to build the device.

After enjoying a lifetime of far-future sightseeing with your time machine - which you kept secret, of course - you get the philanthropic desire to "pay it forward" on your deathbed. You decide to give another intrepid physicist the opportunity to stumble upon this discovery in an exciting way, just as you did. So you draw up some blueprints, lock them in a box, and take one final jaunt in your time machine a century in the past. A few years later, your great-uncle uncovers your buried blueprints ... The causal loop is closed.

Event A (the discovery of the blueprints by your uncle) causes Event B (the production of the time machine), which causes Event C (the burial of the blueprints), which causes Event A ...

Future less Man

Similarly, someone has designed a time machine that goes back a few hours or days. Will he go back and kill if he is killing himself there is now alive? Did you die Who killed him if he died? If he is alive, his future?

Mother or Father?

One of the films about the time travel is Predestination by directors Peter Spierig,

Michael Spierig. A secret agent who travels in a luncheon is working at a hotel. John, who comes there, says about his strange life story, drinking alcohol.

The woman was born John (female name-jane). As soon as he was born, Jane, who was taken to an orphanage by somebody, is growing there. A young child who falls in love after a puberty is born in love with a baby. But that child is missing. Dr. Jen has the right to have two organs in labor during pregnancy. Because of a problem that caused Jane to man. The boy is John.

After telling this story, Jane and her unidentified boyfriend took her to the occasion of the first meeting of Jane's time to find Jane's lover. There, Jane and Jane (both of them are the same!). Love falls. Jane is pregnant. In the end, we know that the agent, Jane, John, and the child are the same person and they are the result of festive encounters with time travels.

Where is the Future Tourists?

Where is the time traveler if it's possible if the trip is possible if it's ever been discovered in the future? Why did not they come yet? If so, is not the trip? If it is possible to make a trip, would our future generations have traveled on time and have already met us?

The first is the **Enrico Fermi** of Italy, so it is also called the **Fermi paradox**.

In practice, it was Stephan Hawking. He made a call for the counter-marketers. He said that "I'm hoping copies of it in one form or another will survive for many thousands of years. Maybe one day someone living in the future will find the information and use a wormhole time machine to come back to my party, proving that time travel will, one day, be possible".

But no one has come as he thought. Many say that the time travel is not possible. But Hawking still now says that the time travel is possible.

TIME TRAVEL IN FICTION

More than scientist film makers and book writers care more important about time travel. In this chapter we seen how the time travel was displayed on stories and films.

The first said, as it was previously seen, H.G. Wells.

Overview

4.   The time traveller will either plan or accident for the past or future.

5.   He uses two different ways to overcome the time. One is shown with the help of

a machine. Another way is by passing through the wormholes known as the portal.

6. The machine is called the time machine. The time machine has been written in different types of stories. It is made up of very small cars, starting with a small hand clock.

7. The portal is the warm hole we've seen before.

8.   A common concept in all the stories is that a new universe develops on all those journeys. That is, new changes are caused by contradictions. That does not make changes to others except for the time travellers.

9.   Most of these stories use Einstein's general theory of relativity to carry out time travels.

10.   A few magical fictional stories called Super Natural. They pass through time

with extraordinary things. For instance, going to the past through the angels.

Time Travel Themes

There are only a few things in these stories and films as the central theme of the story. Let's see what they are,

Changing The Past

The idea of changing the past is logically contradictory, and results in a grandfather paradox. Paul J. Nahin, who has written extensively on the topic of time travel in fiction, states that "*even though the consensus today is that the past cannot be changed, science fiction writers have used the idea of changing the past for good story effect*". Time travel to the past and precognition without the ability to change events may result in causal loops.

The possibilities of deliberate or intentionally changing characters over time led to the idea of "time police", and people worked to avoid such changes by overcoming such changes.

Parallel Universe or Alternate Future

An alternative future or alternate future is a possible future that never comes to pass, typically when someone travels back into the past and alters it so that the events of the alternative future cannot occur, or when a communication from the future to the past effected a change that alters the future. Alternative histories may exist "side by side", with the time traveller actually arriving at different dimensions as he changes time.

The changed future or parallel Universe is the most common. This can change the present time by changing the present time through a message from the future or by transferring it to the past.

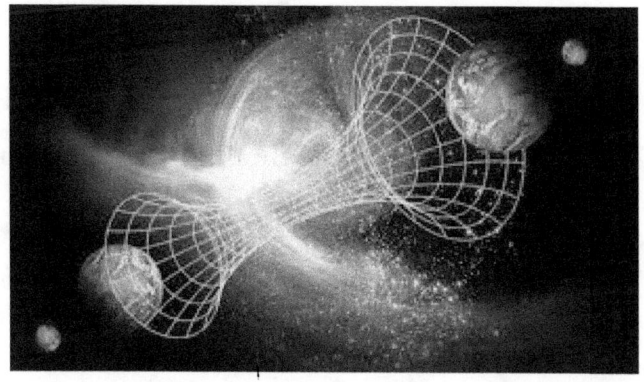

The many stories that come from the future. Let's look at some examples.

11.   The primary example of this is H.G. Wells wrote a short story **The Queer Story of Brownlow's Newspaper** on November 10, 1932, in a November 10, 1972 newspaper. It is the story of the things that take place.

12.   The 1941 film It Happened Tomorrow was written by Robert Sylverberg. In this story, it depicts what we learned from the newspaper. In that story, they found the New York Times on

December 22 on December 1. As the characters know about future events that affect them through a newspaper a week ago, the final outcome is "**stepping up the future of destroying the space**"

13. The film's original TV series Early Edition is created with the embodiment of the film, where the newsletter is available to the protagonist one day earlier and that he uses it.

14. John Buchan's novel **The Gap** in the Curtain presents the same concepts.

Precognition

Precognition has been explored as a form of time travel in fiction. Author J. B. Priestley wrote of it both in fiction and non-fiction, analysing testimonials of precognition and other "temporal anomalies" in his book *Man and Time*. His books include time travel to the future through dreaming, which upon waking up results in memories from the future. Such memories, he writes, may also lead to the feeling of *déjà vu*, that the present events have already been experienced, and are now being re-experienced. Infallible precognition, which describes the future as it truly is, leads to causal loops, a form of which is explored in Newcomb's paradox. The movie *12 Monkeys* heavily deals with themes of predestination and the Cassandra complex, where the protagonist who travels back in time explains that he can't change the past.

Paradox

Many time travel works explore the topic of disrupting causality leading to time paradoxes. One of the most commonly referred to in time travel literature is known as the grandfather paradox. Many works of fiction explore what would happen if a time traveller went back in time and changed the past, for example if they killed their own grandparents.

Time Tourism

A "distinct subgenre" of stories explore the possibility that time travel might be used as a means of tourism, with travelers curious to visit periods or events such as the Victorian Era, Crucifixion of Christ, or some point where dinosaurs could be watched (or hunted by big game hunters), or to meet historical figures such as Abraham Lincoln or Ludwig van Beethoven. This theme can be addressed from two directions. An early example of present-day tourists travelling back to the past is Ray Bradbury's *A Sound of Thunder* (1952), in which the protagonists are big game hunters who travel to the distant past to hunt dinosaurs. An early example of the other type, in which tourists from the future visit our present, is Catherine L. Moore and Henry Kuttner's *Vintage Season* (1946), a story which was selected for inclusion in Volume Two of *The Science Fiction Hall of Fame* collection. In some works, time tourism becomes a problem due to the sheer volume of tourists; for example, in *Up the Line* by Robert Silverberg, "voyeuristic thrill-seekers from the future infest the past".

Time War

Science Fiction describes a time war as a fictional war that is "fought across time, usually with each side knowingly using time travel ... in an attempt to establish the ascendancy of one or another version of history". Time wars are also known as "*change wars*" and "*temporal wars*".

BACK MATTER

Black Hole

A black   hole is   a   region   of space
time exhibiting   such
strong gravitational effects, that nothing—not
even particles and electromagnetic

radiation such   escape from inside it. The general theory of relativity assumes sufficient space for space.

The boundary of the region from which no escape is possible is called the event horizon. Although the event horizon has an enormous effect on the fate and circumstances of an object crossing it, no locally detectable features appear to be observe. In many ways a black hole acts like an ideal black body, as it reflects no light. Moreover, quantum field theory in curved space time predicts that event horizons emit Hawking radiation, with the same spectrum as a black body of a temperature inversely proportional to its mass. This temperature is on the order of billionths of a kelvin for black holes of stellar mass, making it essentially impossible to observe.

Objects whose gravitational fields are too strong for light to escape were first considered in the 18th century by **John Michell** and **Pierre-Simon Laplace.** The first modern solution of general relativity that would characterize a black hole was found by **Karl Schwarzschild** in 1916, although its interpretation as a region of space from which nothing can escape was first published by **David Finkelstein** in 1958. Black holes were long considered a mathematical curiosity; it was during the 1960s that theoretical work showed they were a generic prediction of general relativity. The discovery of neutron stars sparked interest in gravitationally collapsed compact objects as a possible astrophysical reality.

Worm Hole

Wormholes were first theorized in 1916, though that wasn't what they were called at the time. While reviewing another physicist's solution to the equations in Albert Einstein's theory of general relativity, Austrian physicist Ludwig Flamm realized another solution was possible. He described a "white hole," a theoretical time reversal of a black hole. Entrances to both black and white holes could be connected by a space-time conduit.

In 1935, Einstein and physicist Nathan Rosen used the theory of general relativity to elaborate on the idea, proposing the existence of "bridges" through space-time. These bridges connect two different points in space-time, theoretically creating a shortcut that could reduce travel time and distance. The shortcuts came to be called Einstein-Rosen bridges, or wormholes.

Theory of Relativity

The theory of relativity is made by Albert Einstein in two theories: *special relativity and general relativity*. Special relativity applies to the basic particles and their interactions, which illustrate all their physical events except geology. The general relativity explains its relationship with gravity and other forces of nature. This applies to astronomy and astrophysics, including astronomy.

Special Theory of Relativity

Special Theory of Relativity is part of the theory of relativity by Albert Einstein in 1905. It relates to the movement of particles. It says that any movement of the particles is biased and that everything is unresolved. It is noteworthy that in 1687 Newton published policies on the movements. Theoretically travel the following methods.

General Theory of Relativity

This is the *General Theory of Relativity*, published in 1916 by Albert Einstein. This is the theory of gravitation of modern physics. This is a special form of theory and the integral form of Newton's gravitational theory. The general theory of relativity expanded the theory of specialization in order to add gravity. This explains the spatiality of space time caused by mass-energy and dynamic flow. According to the general relativity equations, the universe explains the existence of the "closed periods-like curves", and solutions to these equations that permit the so-called journey of the past.

.

EXTRA MATTER

Albert Einstein

Albert Einstein was a German-born physicist who developed the general theory of relativity. He is considered one of the most influential physicists of the 20th century. Synopsis Born in Ulm, Württemberg, Germany in 1879, Albert Einstein had a passion for inquiry that eventually led him to develop the special and general theories of relativity. In 1921, he won the Nobel Prize for physics for his explanation of the photoelectric effect and immigrated to the U.S. in the following decade after being targeted by the Nazis. Einstein is generally considered the most influential physicist of the 20th century, with his work also having a major impact on the development of atomic energy. With a focus on unified field theory during his later years, Einstein died on April 18, 1955, in Princeton, New Jersey. Miracle Year While working at the patent office, Einstein had the time to further ideas that had taken hold during his studies at Polytechnic and thus cemented his theorems on what would be known as the principle of relativity. In 1905—seen by many as a *"miracle year"* for the theorist—Einstein had four papers published in the Annalen der Physik, one of the best known physics journals of the era. The four papers focused on the photoelectric effect, Brownian motion, the special theory of relativity (the most widely circulated of the write-ups) and the matter/ energy relationship, thus taking physics in an electrifying new direction. In his fourth paper, Einstein came up with the equation $E = mc2$, suggesting that tiny particles of matter could be converted into huge amounts of energy, foreshadowing the development of atomic power. Famed quantum theorist Max Planck backed up the assertions of Einstein, who thus became a star of the lecture circuit and academia, taking on various positions before becoming director of the Kaiser Wilhelm Institute for Physics from 1913 to 1933. Relativity and Nobel Prize In November, 1915, Einstein completed the general theory of relativity, which he considered the culmination of his life research. He was convinced of the merits of general relativity because it allowed for a more accurate prediction of planetary orbits around the sun, which fell short in Isaac

Newton's theory, and for a more expansive, nuanced explanation of how gravitational forces worked. Einstein's assertions were affirmed via observations and measurements by British astronomers Sir Frank Dyson and Sir Arthur Eddington during the 1919 solar eclipse, and thus a global science icon was born. In 1921, Einstein won the Nobel Prize for Physics though he wasn't actually given the award until the following year due to a bureaucratic ruling. Because his ideas on relativity were still considered questionable, he received the prize for his explanation of the photoelectric effect though Einstein still opted to speak about relativity during his acceptance speech. In the development of his general theory, Einstein had held on to the belief that the universe was a fixed, static entity, aka a "cosmological constant," though his later theories directly contradicted this idea and asserted that the universe could be in a state of flux. Astronomer Edwin Hubble deduced that we indeed inhabit an expanding universe, with the two scientists meeting at the Mount Wilson Observatory near Los Angeles in 1930. While Einstein was travelling and speaking internationally, the Nazis, led by Adolf Hitler, were gaining prominence with violent propaganda and vitriol in an impoverished post-WWI Germany. The party influenced other scientists to label Einstein's work "Jewish physics." Jewish citizens were barred from university work and other official jobs, and Einstein himself was targeted to be killed.

Stephen Hawking

Stephen Hawking was born on January 8, 1942, in Oxford, England. At an early age, Hawking showed a passion for science and the sky. At age 21, while studying cosmology at the University of Cambridge, he was diagnosed with amyotrophic lateral sclerosis. Despite his debilitating illness, he has done groundbreaking work in physics and cosmology, and his several books have helped to make science accessible to everyone. Part of his life story was depicted in the 2014 film The Theory of Everything.

Research on Black Holes

Groundbreaking findings from another young cosmologist, Roger Penrose, about the fate of stars and the creation of black holes tapped into Hawking's own fascination with how the universe began. This set him on a career course that reshaped the way the world thinks about black holes and the universe.

While physical control over his body diminished (he'd be forced to use a wheelchair by 1969), the effects of his disease started to slow down. In 1968, a year after the birth of his son Robert, Hawking became a member of the Institute of Astronomy in Cambridge. The next few years were a fruitful time for Hawking. A daughter, Lucy, was born to Stephen and Jane in 1969, while Hawking continued with his research. (A third child, Timothy, arrived 10 years later.) He then published his first book, the highly technical The Large Scale Structure of Space-Time (1973), with G.F.R. Ellis. He also teamed up with Penrose to expand upon his friend's earlier work. In 1974, Hawking's research turned him into a celebrity within the scientific world when he showed that black holes aren't the information vacuums that scientists had thought they were. In simple terms, Hawking demonstrated that matter, in the form of radiation, can escape the gravitational force of a collapsed star. Hawking radiation was born. The announcement sent shock waves of excitement through the scientific world, and put Hawking on a path that's been marked by awards, notoriety and distinguished titles. He was named a fellow of the Royal Society at the age of 32, and later earned the prestigious Albert Einstein Award, among other honors. Teaching stints followed, too. One was at Caltech in Pasadena, California, where Hawking served as visiting professor, making subsequent visits over the years. Another was at Gonville and Caius College in Cambridge.

Hawking made news in 2012 for two very different projects . It was revealed that he had participated in a 2011 trial of a new headband-styled device called the iBrain. The device is designed to "read" the wearer's thoughts by picking up "waves of electrical brain signals," which are then interpreted by a special algorithm, according to an article in The New York Times. This device could be a revolutionary aid to Hawking and others with ALS.

CONCLUSION

Human civilization has tens of thousands of years old. Many innovations are believed to be false and then created. From fire to modern smart tools. It is believed that at one time it was impossible.

At the same time, another thing we need to realize is the superstition. So many comments have been imposed on us. These are the two great powers of the human brain.

Time Travel is not yet discovered, and maybe you can find it again. If they do so, they are good for the development of humanity. It's good if it is not able to find it.

ABOUT MYSELF

It's my second book. It's a just effect of my first book (Science Facts : About The Know Things). I release that book on last march. First few months no one supports my book. At that time I thought book writing is not suitable for me. Then few weeks later. I found real supports, from audience. I wrote that book on English and Tamil. Both sell more or less equal. One person mail me to write elaborate at time travel. It's a part of that book.

Julbiharahamed K